# How to Sparkle at Science Investigations



Monica Huns

We hope you and your class enjoy using this book. Other books in the series include:

**Science titles**

| | |
|---|---|
| How to Sparkle at Assessing Science | 978 1 897675 20 5 |
| How to Sparkle at Science Investigations | 978 1 897675 36 6 |

**Maths titles**

| | |
|---|---|
| How to Sparkle at Addition and Subtraction to 20 | 978 1 897675 28 1 |
| How to Sparkle at Beginning Multiplication and Division | 978 1 897675 30 4 |
| How to Sparkle at Counting to 10 | 978 1 897675 27 4 |
| How to Sparkle at Maths Fun | 978 1 897675 86 1 |
| How to Sparkle at Number Bonds | 978 1 897675 34 2 |

**English titles**

| | |
|---|---|
| How to Sparkle at Alphabet Skills | 978 1 897675 17 5 |
| How to Sparkle at Grammar and Punctuation | 978 1 897675 19 9 |
| How to Sparkle at Nursery Rhymes | 978 1 897675 16 8 |
| How to Sparkle at Phonics | 978 1 897675 14 4 |
| How to Sparkle at Prediction Skills | 978 1 897675 15 1 |
| How to Sparkle at Reading Comprehension | 978 1 897675 44 3 |
| How to Sparkle at Word Level Activities | 978 1 897675 90 8 |
| How to Sparkle at Writing Stories and Poems | 978 1 897675 18 2 |

**Festive title**

| | |
|---|---|
| How to Sparkle at Christmas Time | 978 1 897675 62 5 |

Published by Brilliant Publications
Unit 10, Sparrow Hall Farm, Edlesborough, Dunstable, Bedfordshire LU6 2ES, UK

Email: info@brilliantpublications.co.uk
Website: www.brilliantpublications.co.uk
Tel: 01525 222292



Written by Monica Huns
Illustrated by Lynda Murray
Cover photograph by Martyn Chillmaid

Printed in the UK

Printed ISBN: 978 1 897675 36 6
ebook ISBN: 978 0 85747 041 6

First published in 1999
Reprinted 2000, 2003 and 2009
10 9 8 7 6 5 4

# Contents

# Activity suggestions

**My body** **page 8**
Working in a small group, draw round one child on a large piece of paper. With the help of reference books, ask the children to label as many body parts as possible. The sheet can then be used for individual vocabulary reinforcement and as an assessment activity.

**My Happy and Healthy** **page 9**
Use an enlarged version of Mr Happy and Healthy as an introduction to a class topic on health education. What makes Mr Happy and Healthy smile? Is it food? Water? Exercise? Sleep? People to love him? Alternatively, the sheet could be used at the beginning and end of a topic to chart individual increases in understanding.

**Cats and kittens** **page 10**
As an introduction to this sheet, read some stories that illustrate life-cycles (eg *The Very Hungry Caterpillar*, *Baby Goz*, *You'll Soon Grow Into Them Titch*). The sheet can then be used as an individual activity to join up the families before colouring and labelling them correctly with the aid of dictionaries and reference books.

**I spy** **page 11**
Play a challenging version of Kim's Game with a set of similar objects (eg shells or pebbles) to focus on the visual differences between things. Then give the children a timed 'I spy' challenge to spot and draw something in each circle without moving from their places.

**I hear** **page 12**
Include some IT in this activity by asking small groups of children to tape-record the sounds in different parts of the school (eg the kitchen or the hall). They can then play back the tape, identify the sounds and record them on the sheet.

**I taste** **page 13**
The game on this sheet makes a fun class activity when learning about the senses. Mixing two tastes, eg sugar and tomato or apple and marmite, adds extra challenge. (Health and safety notes: using this sheet provides a good opportunity to remind children of the importance of washing their hands before touching food. You will need to be aware whether any of the children have any food allergies.)

**I feel** **page 14**
Using a blindfold or feely box, play games that focus on vocabulary connected with touch (eg 'Can you find me something cold and hard in the box?'). Then take the children on a scavenger hunt in the school grounds to collect as many different feely things as possible, eg grit, leaves, feathers. Sticking the bits on Charlie Caterpillar provides a way of recording what the chilldren found on the hunt and of assessing their understanding of vocabulary.

**Sniffer dog auditions** **page 15**
Introduce the sheet by playing a class detective game. Explain that someone has dropped the coffee mug and Sniffer Dog Detective will be able to guess who by the smell of coffee on his/her tissue. Choose one child to be Sniffer Dog Detective. While he/she is out of the room give each child a crumpled up tissue, one of which has been dipped in coffee. Then when he/she returns, Sniffer Dog Detective must find out the culprit by sniffing the tissues. Using the sheet, small groups of children can then carry on the auditions on their own.

**What do plants need to grow?** **page 16**
This sheet can be used to record a class investigation or in smaller groups as a means of assessing individual Sc1 skills. Use small bedding plants, such as lobelia, and involve the children in discussion about what they want to find out and where they should therefore put the plants. Fill in the 'Where?' and 'What do you think will happen?' parts at the beginning of the investigation. Add the observational drawing, the 'When?' and 'Were you right?' at the end. With some additional questioning (as appropriate) this sheet will support a sound assessment of individual progress in science.

**A flowering plant** **page 17**
Look at a flowering plant together and ask the children to identify as many parts as possible. Do they think every plant will have these parts? Then, working in groups with a variety of flowering plants, let everyone fill in their sheet individually. Share the sheets at the end of the lesson and discuss the different parts labelled.

**Matching seeds** **page 18**

Prepare this group activity by slicing the six fruits in half and removing a few seeds from each. Give each child some seeds to match with the correct fruit, using lenses to encourage close observation. Finally, ask them to carefully sketch the inside of the fruit and stick their seeds in the appropriate place on the drawing. Leftover seeds can be washed and planted in small pots of compost to see if they will grow.

**Same or different?** **page 19**

Introduce the idea of similarities and differences by playing some circle guessing games, eg 'Who touched me?' (when one blindfolded child is gently touched by another and has to guess who it was) or 'Tomato sauce' (when one child looks away and has to guess who said 'tomato sauce'). Ask the children to work in pairs to fill in the sheet, looking for similarities and differences. The inclusion of height, weight and handspan adds opportunities for measuring.

**A leaf sort** **page 20**

It might be useful to enlarge the sheet before providing the children with a range of leaves. Encourage them to observe the leaves closely with a lens and to sort the leaves into groups, eg by colour, by shape, by arrangement on the stem, before sticking them on to the page. A similar format sheet with petal or seed shapes would provide further practice in grouping living things.

**Where do mini-beasts live?** **page 21**

This investigation will need to be set up at least one week before the children begin to fill in the sheet. Place an upturned flower pot, some rotting wood (or a piece of carpet) and a couple of bricks or stones in a corner of the school grounds, preferably on soil or grass. Without disturbing the habitats too much, the children can then draw the mini-beasts they find each week. Use reference books back in the classroom to help with identification. By studying the results at the end of the three weeks the children can be encouraged to raise further questions of their own about the mini-beasts, eg 'Why are there more woodlice under the dead wood than under the flower pot?'

**Sorting fabrics** **page 22**

Introduce the children to the idea of a patchwork quilt by showing them an example or by reading a story, such as *The Patchwork Cat*. Then give them a variety of prepared pieces of fabric to make their own selection to stick on the sheet. They will be using their observational skills and their understanding of the vocabulary to complete the task.

**Is it see-through?** **page 23**

This is a good activity for encouraging independent investigation. Start the children off in pairs with an array of interesting objects to shine a torch through, including some different fabrics and a range of papers. They can then move on to investigating and classifying other materials used in the classroom.

**What will the magnet pick up?** **page 24**

Spread out the objects to be tested. Before giving the children magnets, ask them to predict which will go in the magnetic hoop and which in the non-magnetic hoop. This will reveal not only their understanding, but any misconceptions that may have, eg all metals are magnetic. When they have tested the objects with a magnet, ask them if their results have changed any of their ideas.

**Which magnet is the strongest?** **page 25**

Give pairs of children a range of magnets, a large pile of paper-clips and some squared paper. Stick this question to the table: 'Which magnet is the strongest?' Discuss with the group a variety of ways of tackling the investigation before each pair gets to work on their own. Some might want to count how many paper-clips each magnet will pick up; some might want to count from how many squares away a paper-clip is attracted to a magnet; some pairs might see the value in checking their results by investigating the same question in two different ways. Observing and questioning them as they work on this sheet provides a good way of assessing their investigative skills.

**We're going on a materials hunt** **page 26**

This could be a useful introduction to a topic on materials or an end-of-topic assessment task. Go exploring around the outside of the school buildings to fill in the snake, either by colouring in each section, or by drawing and labelling something appropriate in each part. Back in the classroom, use the snakes as a basis for a discussion on why different parts of the building are made of different materials, eg glass for windows, clay tiles for the roof.

**Which hat for teddy?** **page 27**

This investigation requires a teddy and three hats. One hat should be completely waterproof; the others might be paper, wool, straw or some other fabric. Using a pipette for raindrops introduces the children to simple scientific equipment, and the final question on the sheet encourages them to make simple comparisons.

**A nice hot cup of tea** **page 28**
This activity provides an opportunity for the children to practice reading a dip thermometer. The children could practice testing a range of drinks (eg iced water, milk from a carton) whilst the teapot is cooling to a safe temperature. Then, working in pairs or small groups, ask them to insulate their cups with cotton wool, foil and newspaper before pouring the tea. They can use the sheet to record the temperature of the liquid straight after pouring and half an hour later. This investigation provides the opportunity to raise issues concerning a fair test (eg Are all the insulation layers the same thickness? Has each cup got an equal volume of tea?).

**Making models** **page 29**
This sheet makes explicit the scientific principles behind an everyday activity in the infant classroom. It also reinforces the use of appropriate vocabulary. Can the children think of any other materials that can be changed in these ways?

**All change?** **page 30**
This is a useful hands-on activity to get the children thinking about the effects of heat. The idea can then be reinforced and extended by a variety of activities, such as baking with chocolate, candle-making and ice balloon play in the water tray (see page 33).

**Changing for good?** **page 31**
This sheet provides another opportunity to draw out the science from an everyday infant activity. Can the children think of anything else that is changed for good by heating it?

**The wise man and the foolish man** **page 32**
Introduce this activity by telling the children the biblical story of the wise man and the foolish man (Matthew 7.24-27). Ask for their suggestions for testing the scientific truth of the story before giving a class demonstration or letting them have a go at a test on their own. The last question on the sheet directs attention to issues of fair testing (eg How much water did you use each time? Did you build two identical houses?). This experiment also makes a good focal point for an assembly.

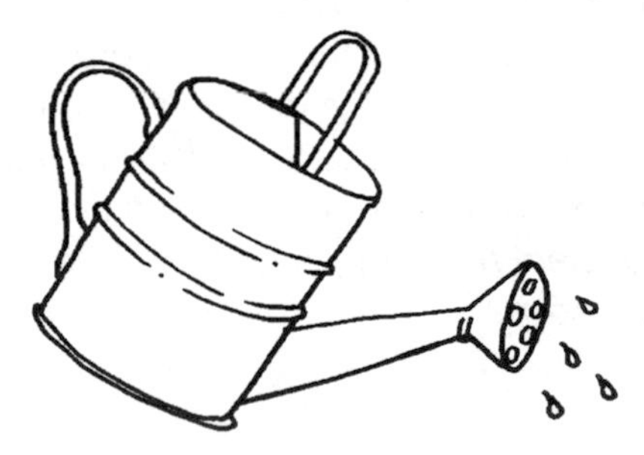

**Ice balloons** **page 33**
Prepare some ice balloons by filling balloons with water and leaving them in the freezer for a few days. Then give small groups of children a balloon to investigate with a range of measuring equipment (eg scales, thermometer, and a tank of water). After an appropriate interval introduce some salt and food colourings for further exploration. The sheet provides a framework for recording some of the children's discoveries. It can easily be filled in at a later date, leaving the children free, initially, to concentrate on the investigative process.

**Testing the bridge** **page 34**
The focus of this investigation is on making a prediction and then considering whether the evidence supports that prediction or not. Ask the children to fill in the prediction bubble before carrying out the test. The bridges can be made with two towers of bricks or heavy books and identically sized strips of wood, paper, plastic and cardboard. At the end of the lesson a group presentation to the rest of the class provides an opportunity to discuss the accuracy of the predictions.

**Electricity in the home** **page 35**
This sheet focuses on the possible dangers of electricity in the home, but it also serves to highlight its many uses. It could be used as a homework task, perhaps asking the children to draw a safe version of the picture. (Safety note: remind the children never to play with electricity as it is very dangerous.)

**Making a circuit** **page 36**
Give a group of children a selection of the items shown on this sheet for investigative play. How many different working circuits can they make? The sheet can then be used as a record of their activity and an assessment of their understanding of the principles of circuit making (eg Is there a battery included? Is the circuit continuous?).

**Which vehicle goes furthest?** **page 37**
This investigation focuses on predicting which vehicle will travel furthest after being rolled down a ramp. A period of exploratory play is therefore important before the prediction is made with some adult questioning to encourage the children to look at the characteristics of five or six toy cars (eg wheel size, materials used, mass). A plank of wood or a piece of PE equipment can then be used to carry out the test. Measuring the distance each car travels will extend the scope of this activity and gives some elements of Sc1.

**Moving toys** **page 38**

Assemble a collection of moving toys for investigative play. You might include windmills, wind-ups, a yo-yo, some battery-operated cars, a skipping rope. When the children have had lots of opportunity to explore, ask them to choose a favourite toy to draw. Adding the arrows challenges them to show understanding of the forces (eg pushes or pulls) which make the toy move.

**Investigating sails** **page 39**

This investigation combines the exploration of forces with an opportunity to discuss fair testing. Three identical boats and a large water tray are required, but the investigation will also work with three identical cars. First the children need to make their different shaped sails on straw masts and attach them with Blu-tack to the boats. Then they need to discuss how to test the sails' performance fairly before making a prediction. Should, for example, the same person blow each time? Should all the boats start from the same place? After the test, the sheet can be used for individual recording and as the basis for teaching, questioning and assessing individual understanding.

**Investigating with forces** **page 40**

This investigation works best if each object is rolled down a bowling alley made from two planks or pieces of PE equipment. The children can then observe what happens when the object meets the air from the hair dryer or the water from the hose (or washing-up liquid bottle). The sheet introduces the idea of recording results in table form with the opportunity to compare the force of the air with the force of the water.

**Floating and sinking** **page 41**

With a tank and a collection of objects, this sheet can be used to support children in some independent investigating. Before doing the investigation, ask the children to circle the things they think will float. At the end of the activity ask if they have had any surprises. Their results can form the basis for a class discussion on why some things float and some things sink. You could introduce the significance of something being heavy for its size by looking at the middle set of objects.

**Boats and cargo** **page 42**

This activity is a useful follow-up to Floating and sinking (page 41). It enables children to put into practice what they have understood about why some things float and some things sink. If appropriate, the teacher can then introduce the concept of upthrust as part of the balance of forces which keeps things afloat.

**Light sources** **page 43**

A treasure hunt to search for all the light sources in school (or a homework task to draw all the light sources at home) makes a good introduction to this topic. After discussion this sheet can then be used to reinforce the children's learning and provide assessment evidence for the teacher.

**Shadows** **page 44**

Using a projector or angle poise lamp and a blank wall, give the children plenty of opportunity for investigative play with shadows. Then use the sheet for some closer observation by shining a light on to the five objects arranged on a large piece of white paper. What shapes are the shadows? Are they equally dark all over? Providing a range of pencils from B to 6B will help the children to draw and shade the shadows accurately.

**Colours in the dark** **page 45**

Line a shoe box with black paper so that the children can carry out this investigation, working in pairs. The cars should include one white or yellow and one dark blue or green. The results of the experiment could contribute to a discussion on road safety and a technology task to design a waistcoat for safer cycling or walking at night.

**Sound walk** **page 46**

This activity makes a good introduction to a topic on sound. Even a short walk will provide plenty of ideas for the children to draw on their sheets back in the classroom. Using bubbles for the sounds is fun and can be linked to punctuation work on speech marks.

**Making sounds** **page 47**

This sorting task encourages the children to look more carefully at how sounds are made. It can be done by giving a group of children three hoops and any combination of instruments available in school to play and sort. They can then use the sheet to record their findings. This activity leads on to an explanation of how sounds are only heard when the vibrations affect the small bone in the ear.

**Investigating sound** **page 48**

This investigation works well as a class demonstration with one child being tested wearing a woolly hat and another child without. Measure the furthest distance from which the sounds can still be heard and then compare the two sets of results. The sheet will support small groups of children testing their own hearing and comparing results.

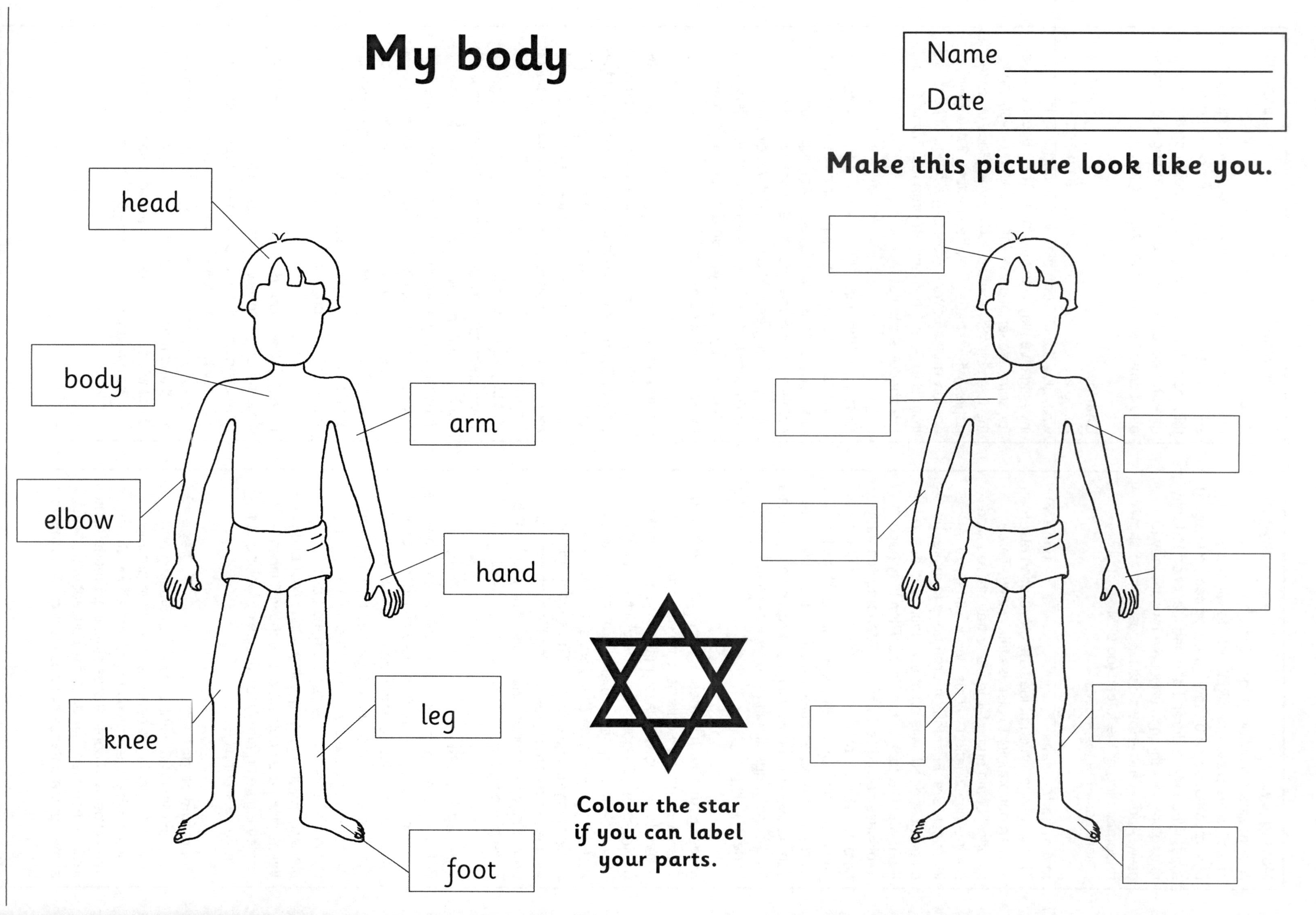
My body
Name
Date
Make this picture look like you.
head
body
arm
elbow
hand
knee
leg
foot
Colour the star
if you can label
your parts.

# Mr Happy and Healthy

Name ______________________

Date ______________________

**Write or draw things you can do to keep happy and healthy.**

**Colour the star if you know seven ways to stay happy and healthy.**

# Cats and kittens

Name ____________

Date ____________

**Draw lines to match the babies to their mothers.**

**Word box**

| | | | |
|---|---|---|---|
| cat | kitten | butterfly | caterpillar |
| hen | chick | duck | duckling |
| sheep | lamb | dog | puppy |
| frog | tadpole | woman | child |

**Colour the star if you can label all the pictures.**

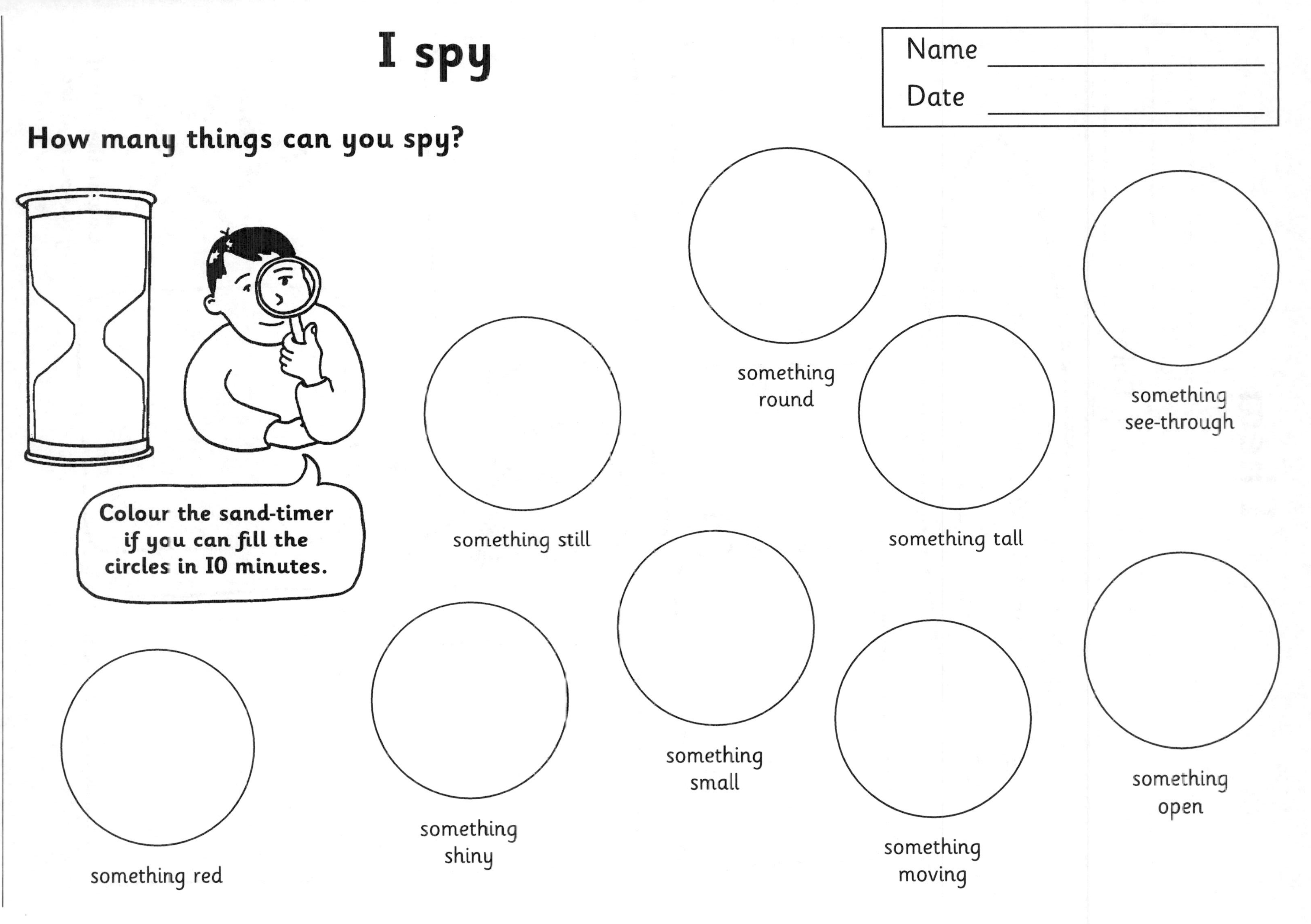
I spy
Name
Date
How many things can you spy?
Colour the sand-timer if you can fill the circles in 10 minutes.
something still
something round
something tall
something see-through
something red
something shiny
something small
something moving
something open

# I hear

Name ______________________

Date ______________________

These are the sounds I heard in

______________________________

**Colour the star if you heard at least six sounds.**

# I taste

Name ____________

Date ____________

Put small bits of different foods in the pots.
Now blindfold some friends and play a tasting game.
Put a tick ✓ for a right guess.
Put a cross ✗ for a wrong one.

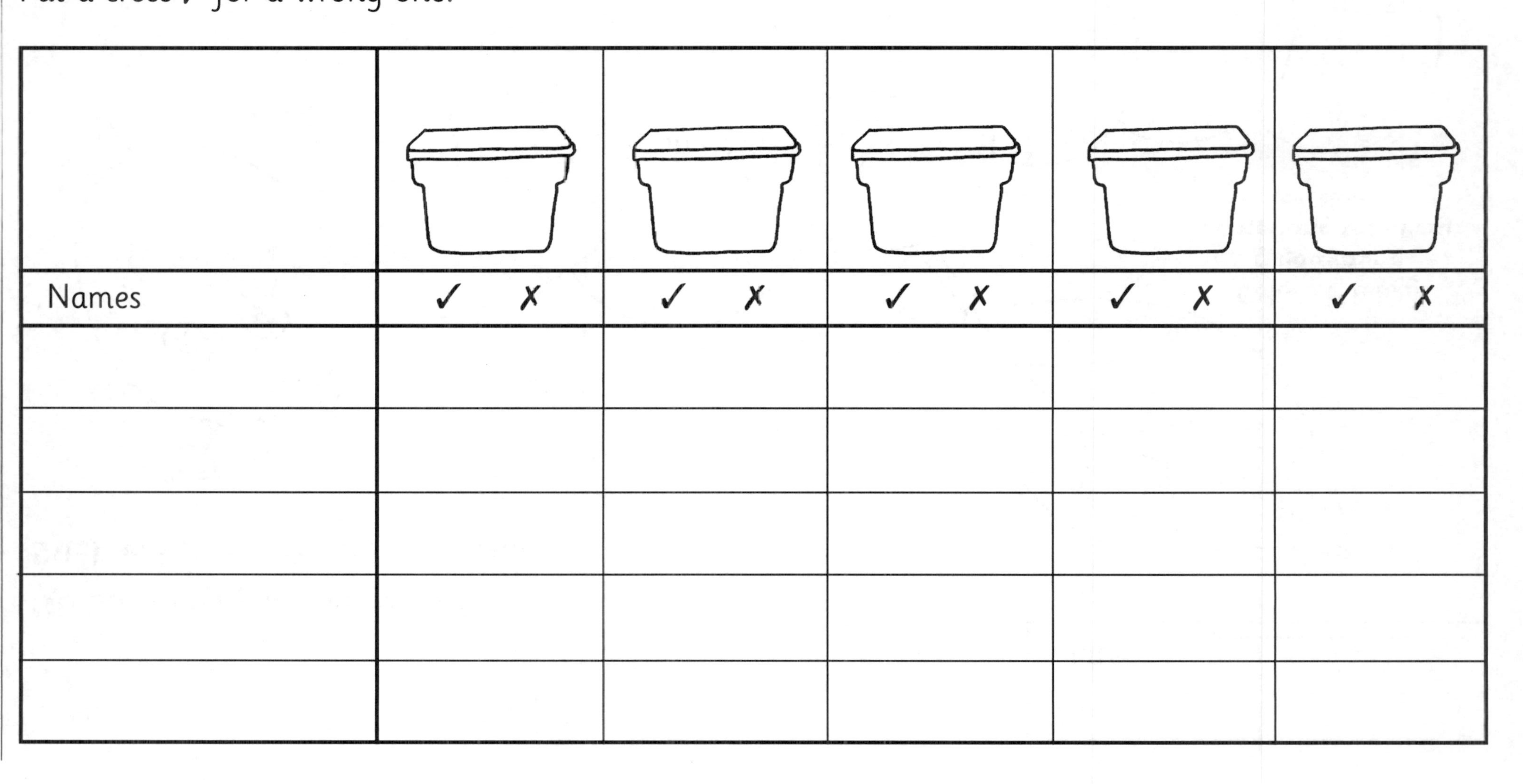

| | | | | |
|---|---|---|---|---|
| Names | ✓ ✗ | ✓ ✗ | ✓ ✗ | ✓ ✗ | ✓ ✗ |
| | | | | | |
| | | | | | |
| | | | | | |
| | | | | | |
| | | | | | |

# I feel

Name ____________________

Date ____________________

**Go on a feely scavenger hunt.**

**Stick what you find on Charlie.**

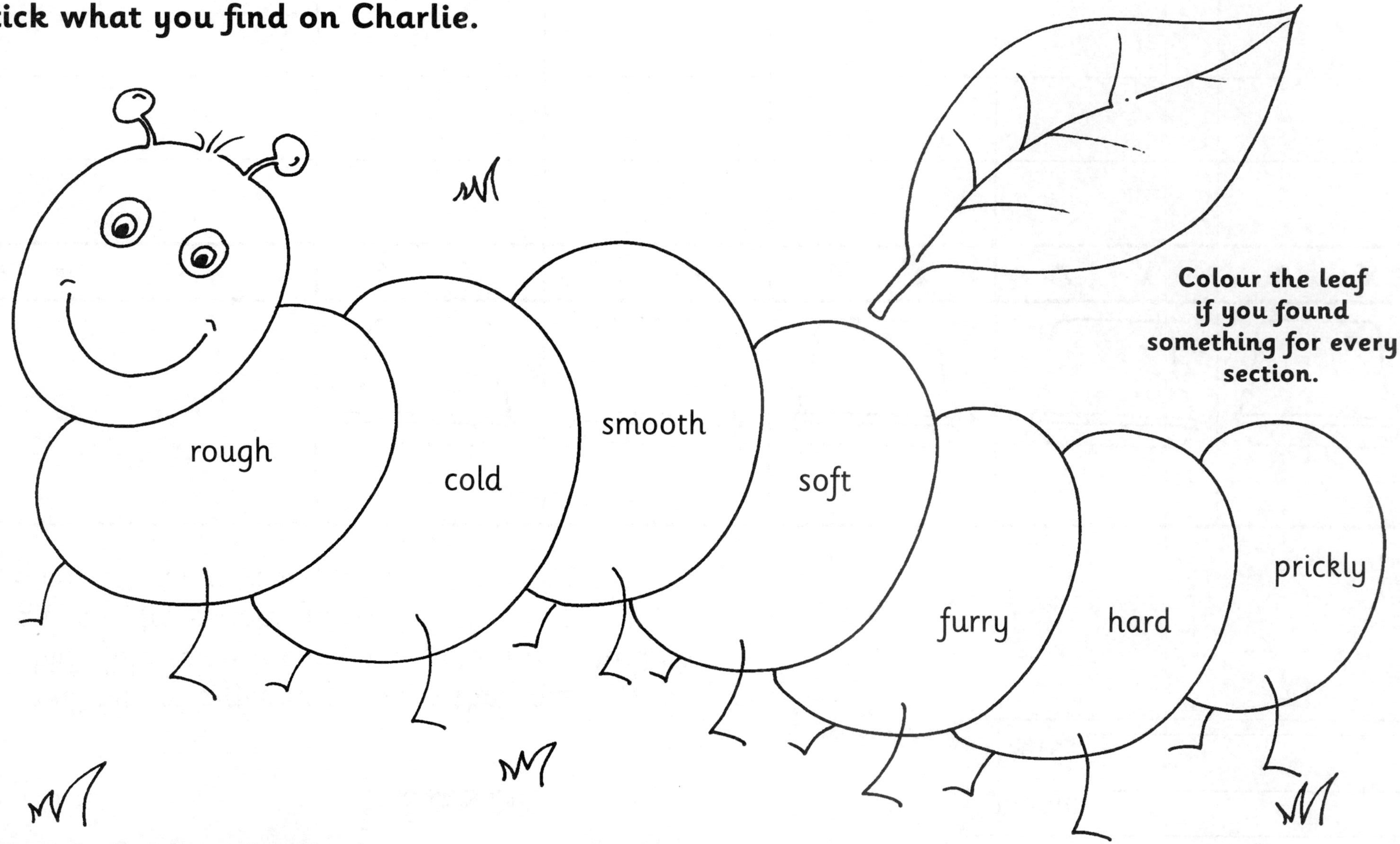

rough

cold

smooth

soft

furry

hard

prickly

**Colour the leaf if you found something for every section.**

# Sniffer dog auditions

Name ______________________

Date ______________________

Drop a different smell on to each tissue. Now blindfold some friends and see if they can guess the smell.
Put a tick ✓ for a right guess.
Put a cross ✗ for a wrong one.

**I think ______________________**

**would make the best Sniffer Dog Detective.**

| Names | ✓ | ✗ | ✓ | ✗ | ✓ | ✗ | ✓ | ✗ | ✓ | ✗ |
|---|---|---|---|---|---|---|---|---|---|---|
| | | | | | | | | | | |
| | | | | | | | | | | |
| | | | | | | | | | | |
| | | | | | | | | | | |
| | | | | | | | | | | |

# What do plants need to grow?

Name ______________

Date ______________

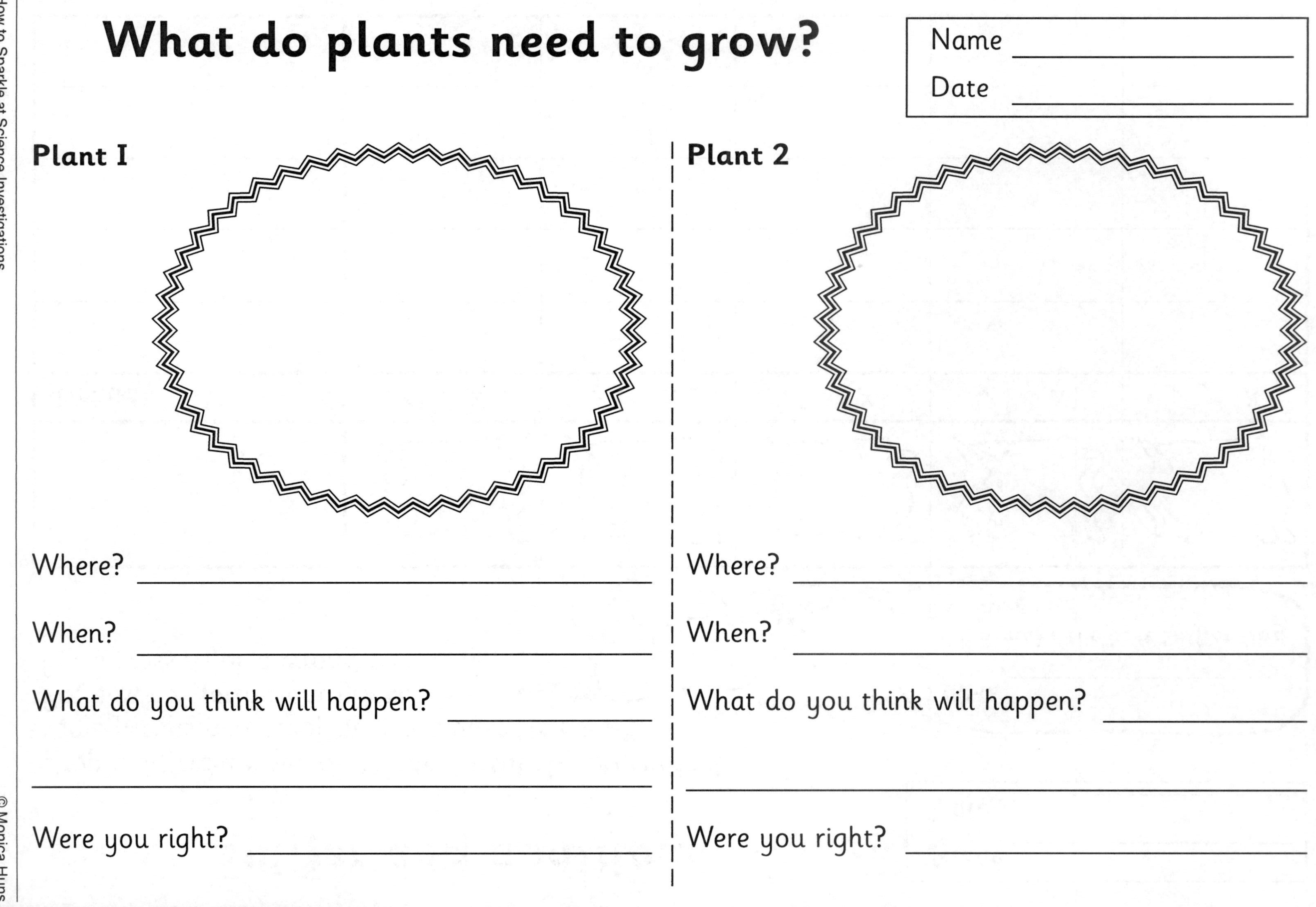

**Plant I**

Where? ______________

When? ______________

What do you think will happen? ______________

______________

Were you right? ______________

**Plant 2**

Where? ______________

When? ______________

What do you think will happen? ______________

______________

Were you right? ______________

# A flowering plant

Name ______________________

Date ______________________

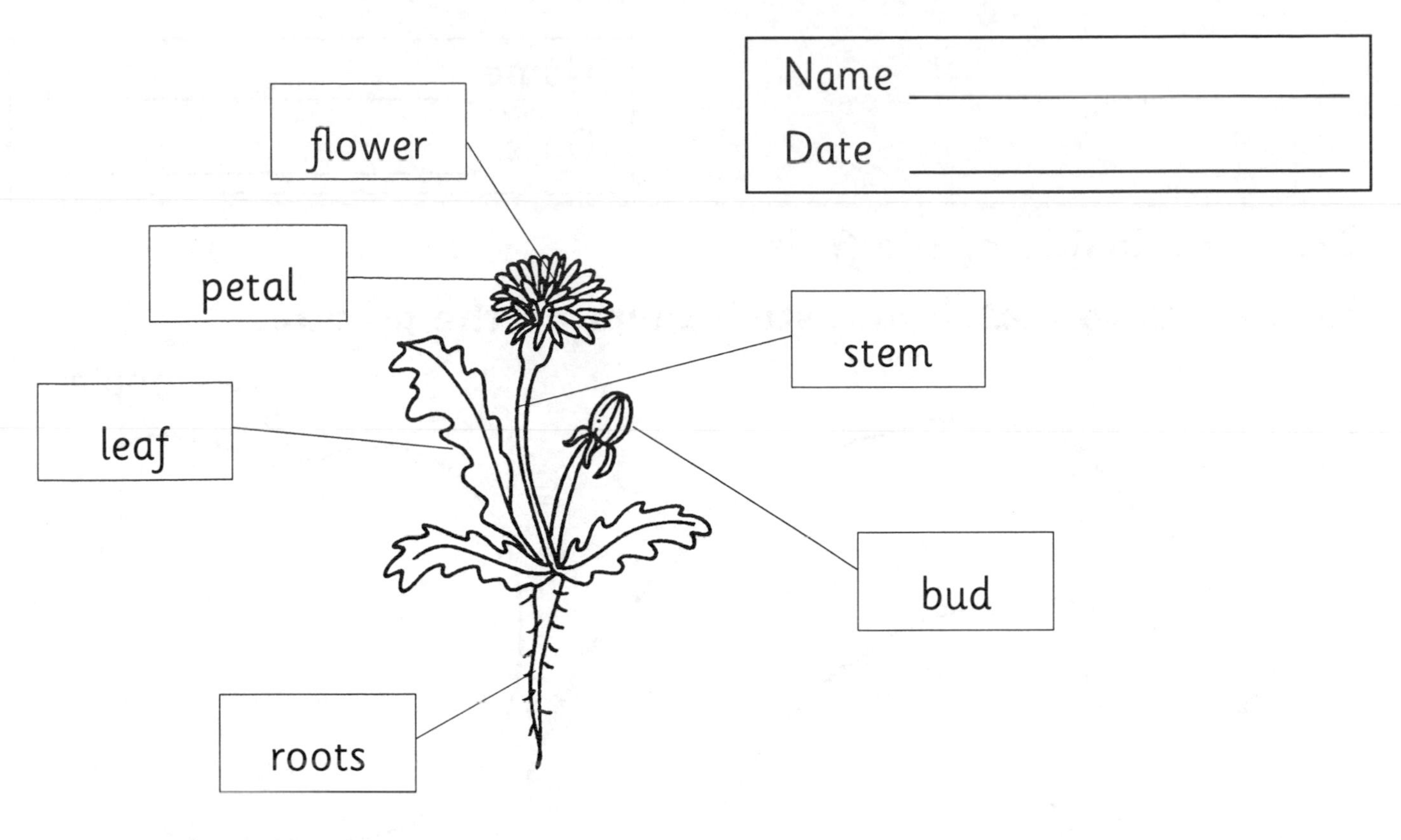

# Matching seeds

Name ____________________

Date ____________________

**Draw the inside of the fruit.**

**Find seeds to match and stick them on the picture.**

# Same or different?

Name ____________________

Date ____________________

**Me**

First draw your face and your friend's face. Then fill in the table.

**My friend**

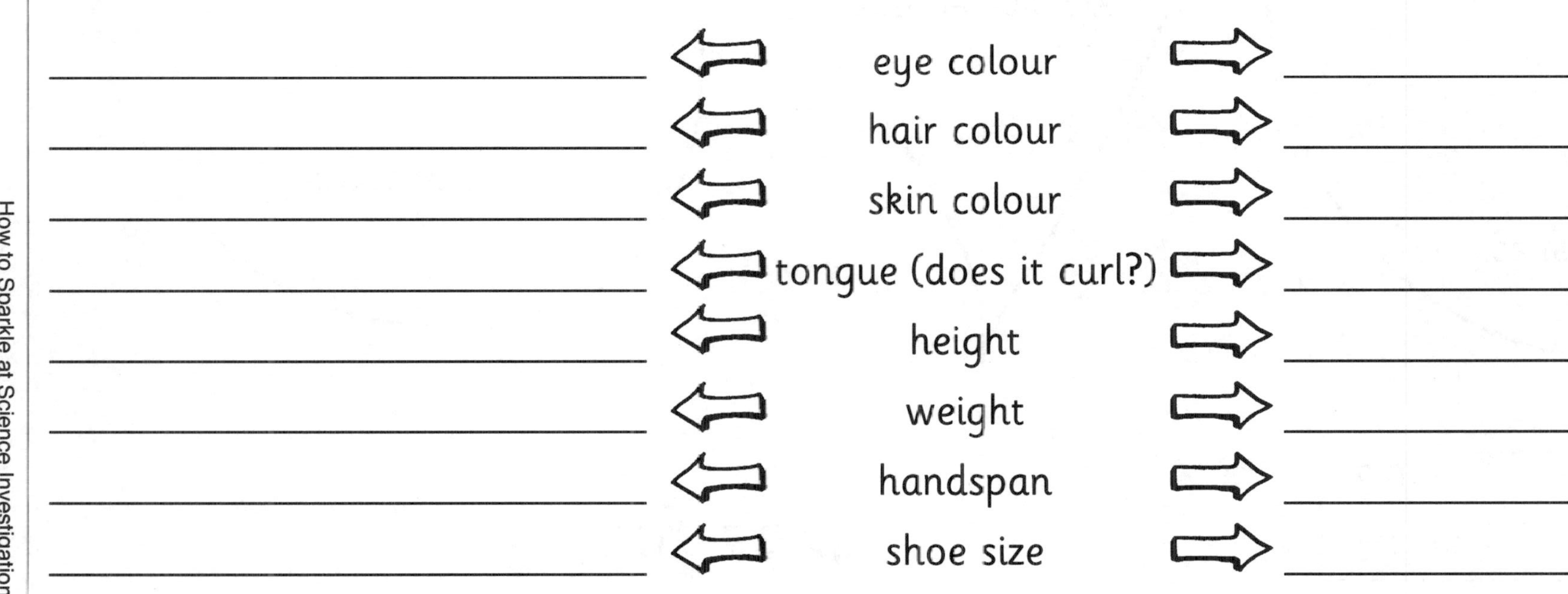

| Me | | My friend |
| --- | --- | --- |
| | eye colour | |
| | hair colour | |
| | skin colour | |
| | tongue (does it curl?) | |
| | height | |
| | weight | |
| | handspan | |
| | shoe size | |

# A leaf sort

Name ____________

Date ____________

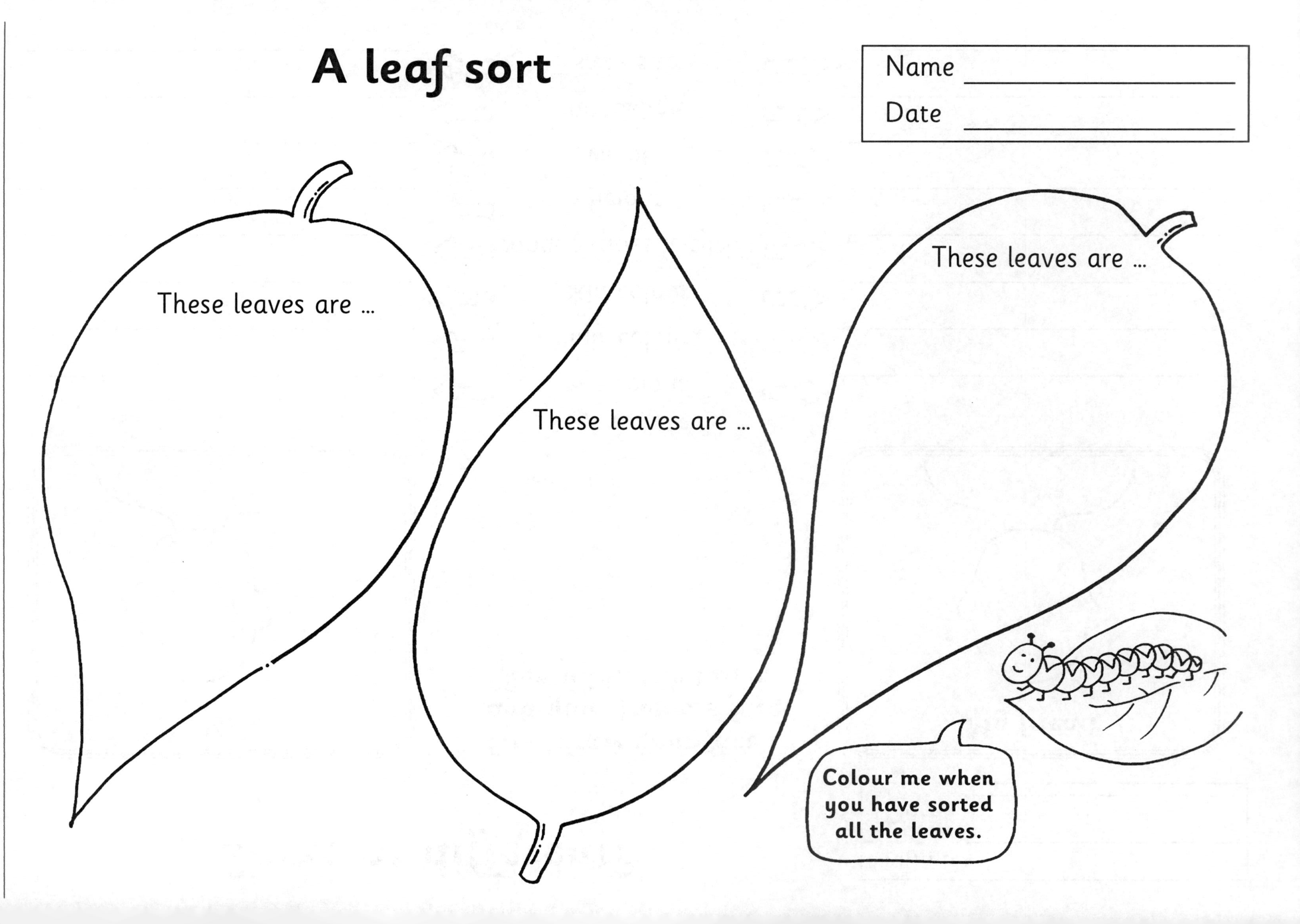

# Where do mini-beasts live?

Name ______________________

Date ______________________

**Look carefully underneath each of these places every week. Draw what you see.**

| | | | |
|---|---|---|---|
| Week I | | | |
| Week 2 | | | |
| Week 3 | | | |

# Sorting fabrics

Name ____________________

Date ____________________

**I can make a patchwork quilt.**

# Is it see-through?

Name ______________________

Date ______________________

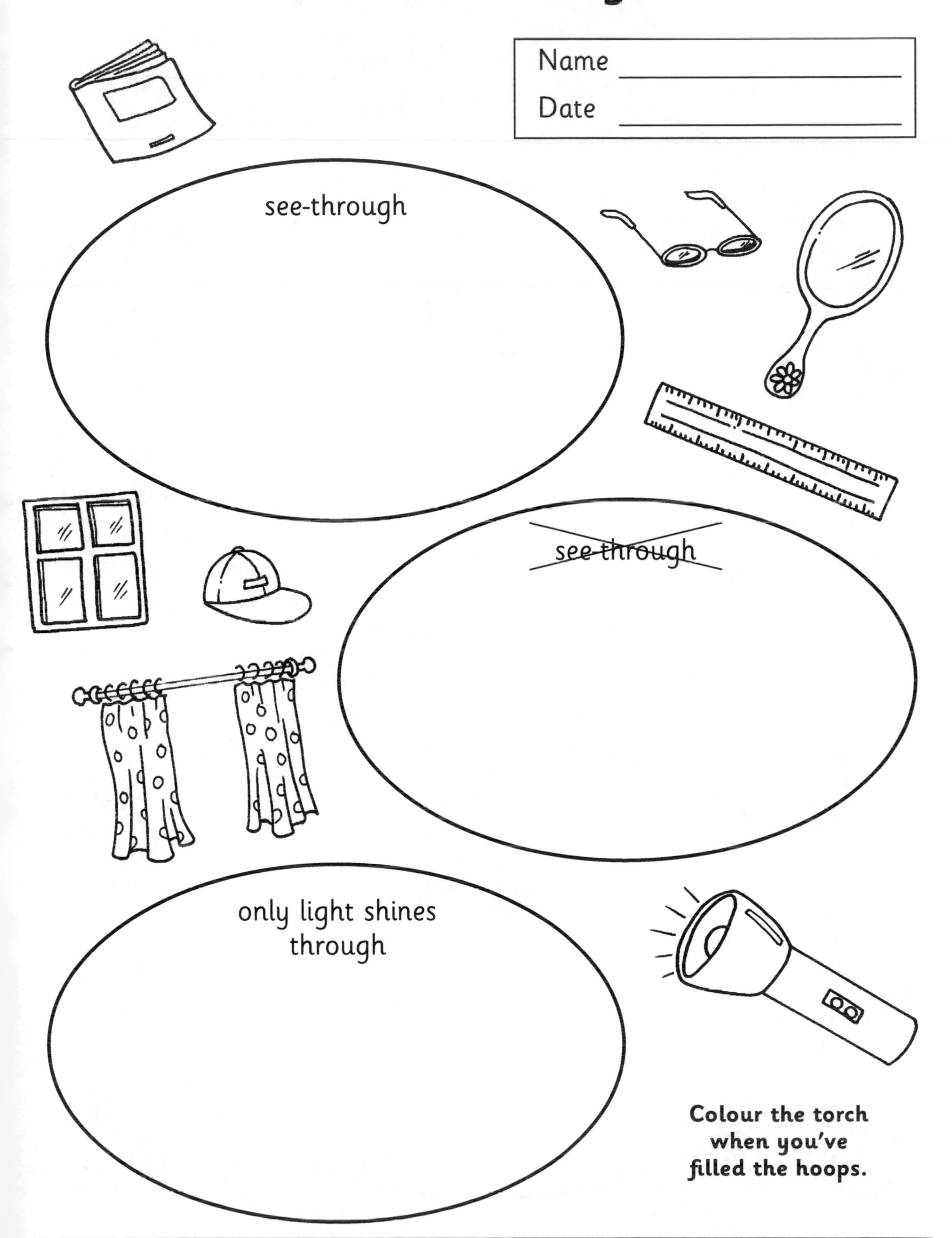


**Colour the torch when you've filled the hoops.**

# What will the magnet pick up?

Name ______________________

Date ______________________

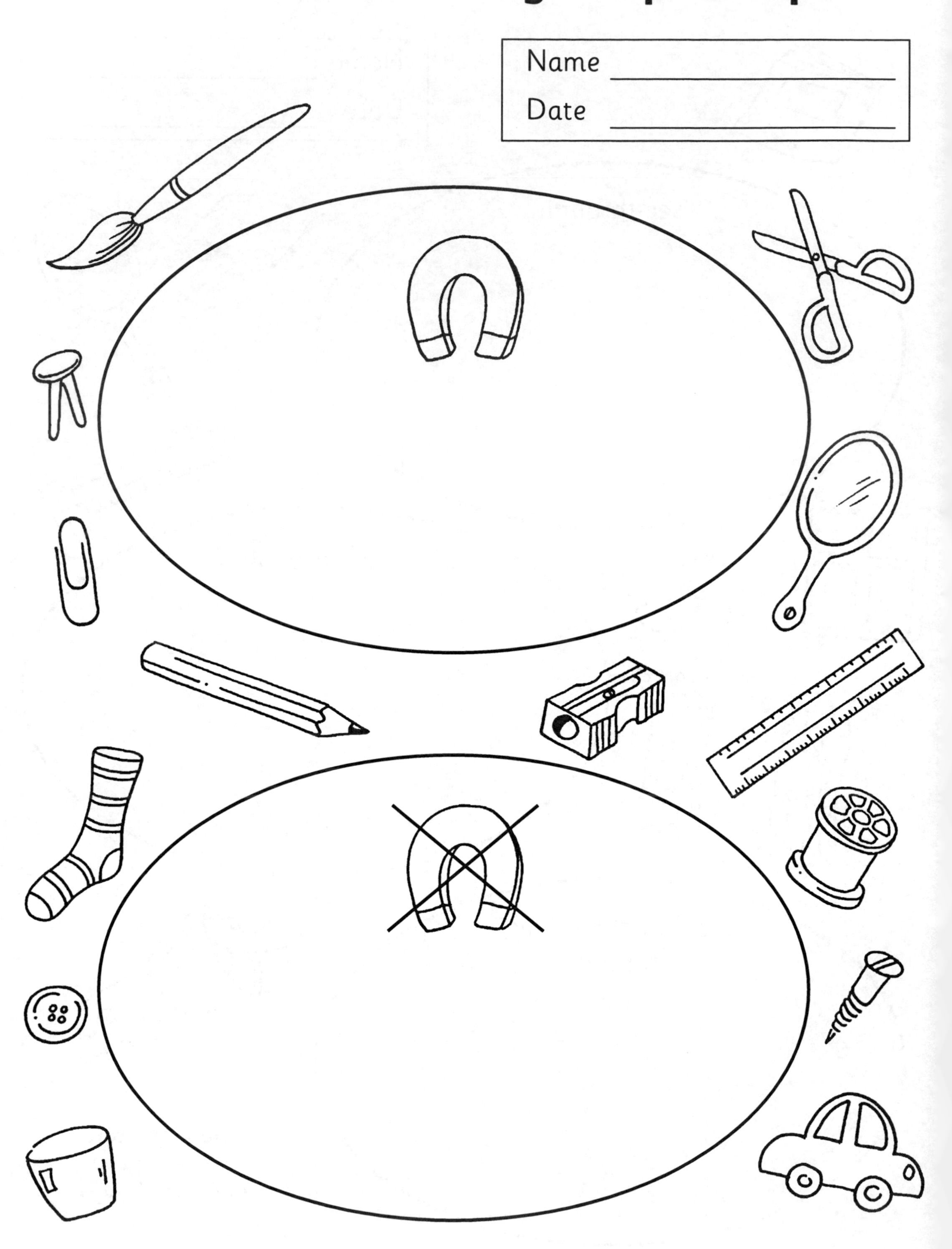

# Which magnet is the strongest?

Name ______________________

Date ______________________

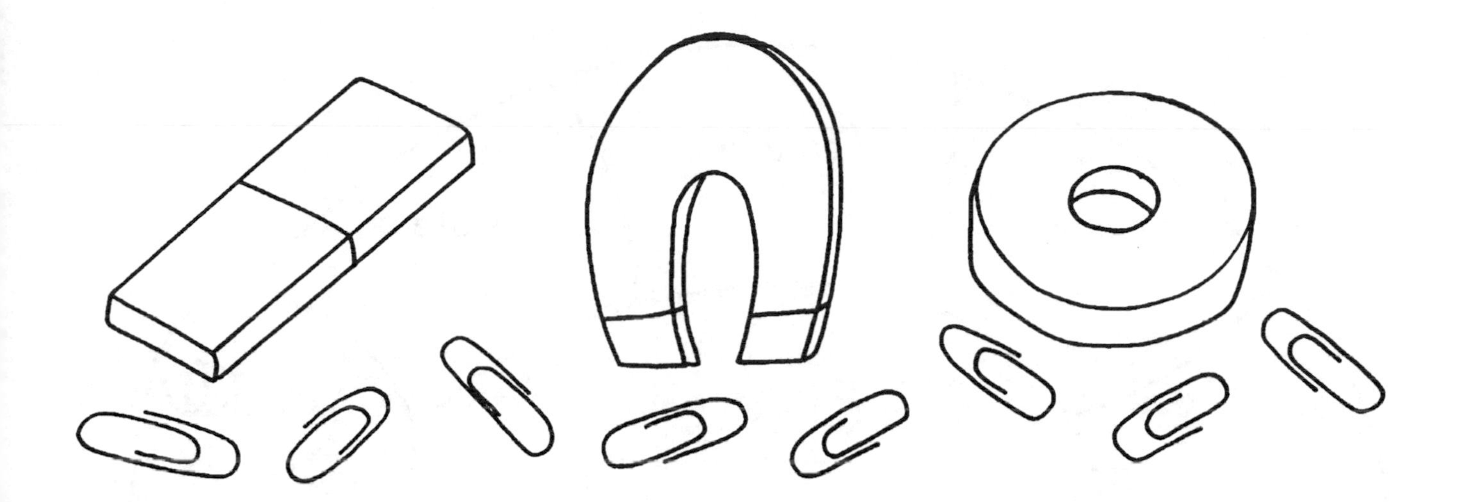

What we did.

What we found out.

# We're going on a materials hunt

Name ______________________

Date ______________________

**Can you spot something for each part of the snake?**

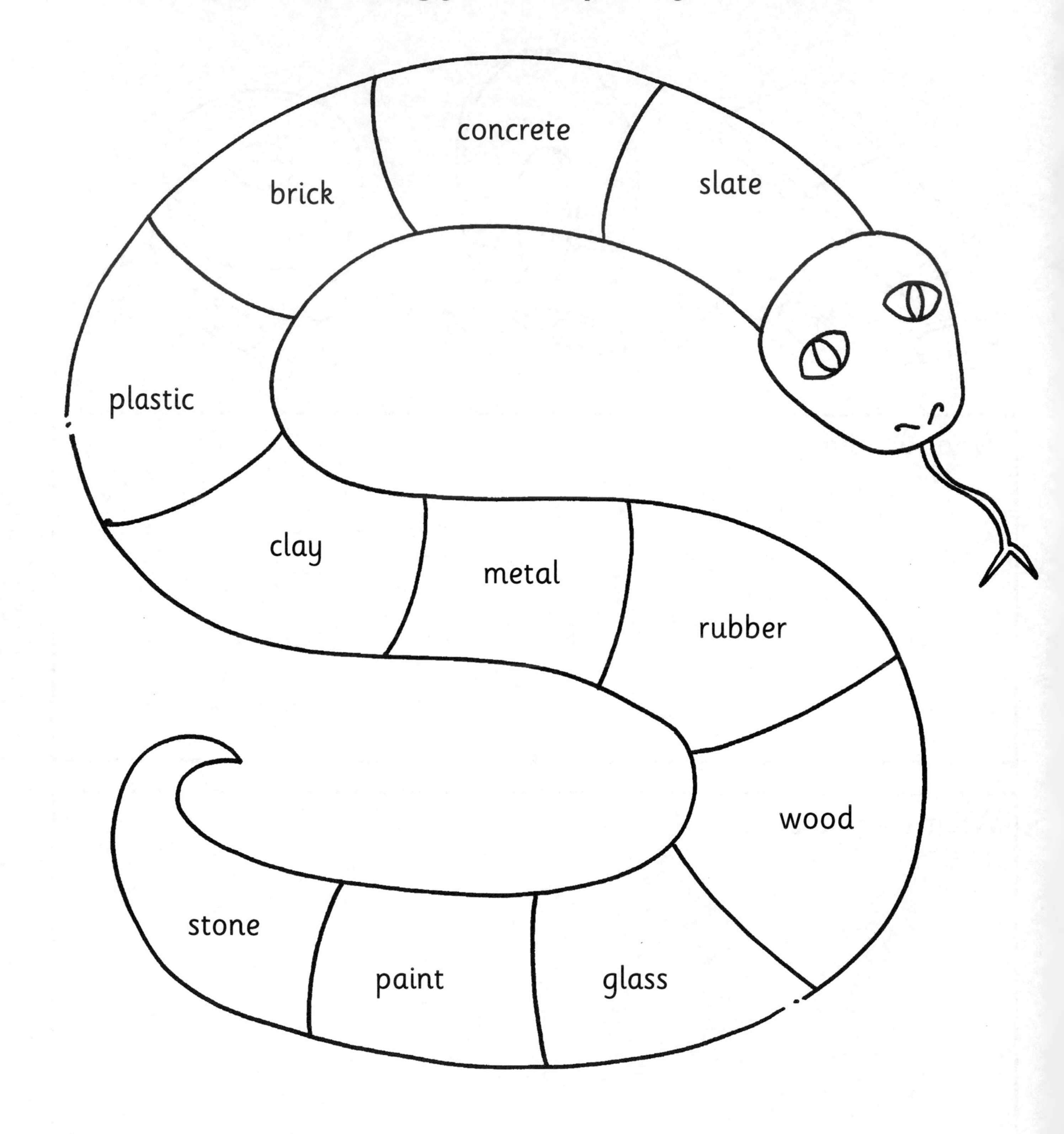

# Which hat for teddy?

Name ______________________

Date ______________________

**Put three different hats on teddy.**

**Use a pipette to make it rain.**

| **Hat 1**<br>How many drops before teddy got wet? | **Hat 2**<br>How many drops before teddy got wet? | **Hat 3**<br>How many drops before teddy got wet? |
|---|---|---|
| | | |

Which hat should teddy wear in the rain?

# A nice hot cup of tea

| Name | ____________________ |
|---|---|
| Date | ____________________ |

**Choose three different covers for the cups.**

**Use a dip thermometer to measure how warm the tea is.**

| **Cover 1** | **Cover 2** | **Cover 3** |
|---|---|---|
| At ____________<br>the tea was ___°C. | At ____________<br>the tea was ___°C. | At ____________<br>the tea was ___°C. |
| At ____________<br>the tea was ___°C. | At ____________<br>the tea was ___°C. | At ____________<br>the tea was ___°C. |

Which cover kept the tea warmest?

# Making models

| Name | |
|---|---|
| Date | |

**Draw four different things you have made with Plasticine, clay, cardboard or Lego.**

| | Put a ✓ if it will: | |
|---|---|---|
| | squash<br>bend<br>twist<br>stretch | |
| | Put a ✓ if it will: | |
| | squash<br>bend<br>twist<br>stretch | |
| | Put a ✓ if it will: | |
| | squash<br>bend<br>twist<br>stretch | |
| | Put a ✓ if it will: | |
| | squash<br>bend<br>twist<br>stretch | |

# All change?

Name ______________________

Date ______________________

**Hold each thing in your hand for two minutes.**

**Draw what happened.**

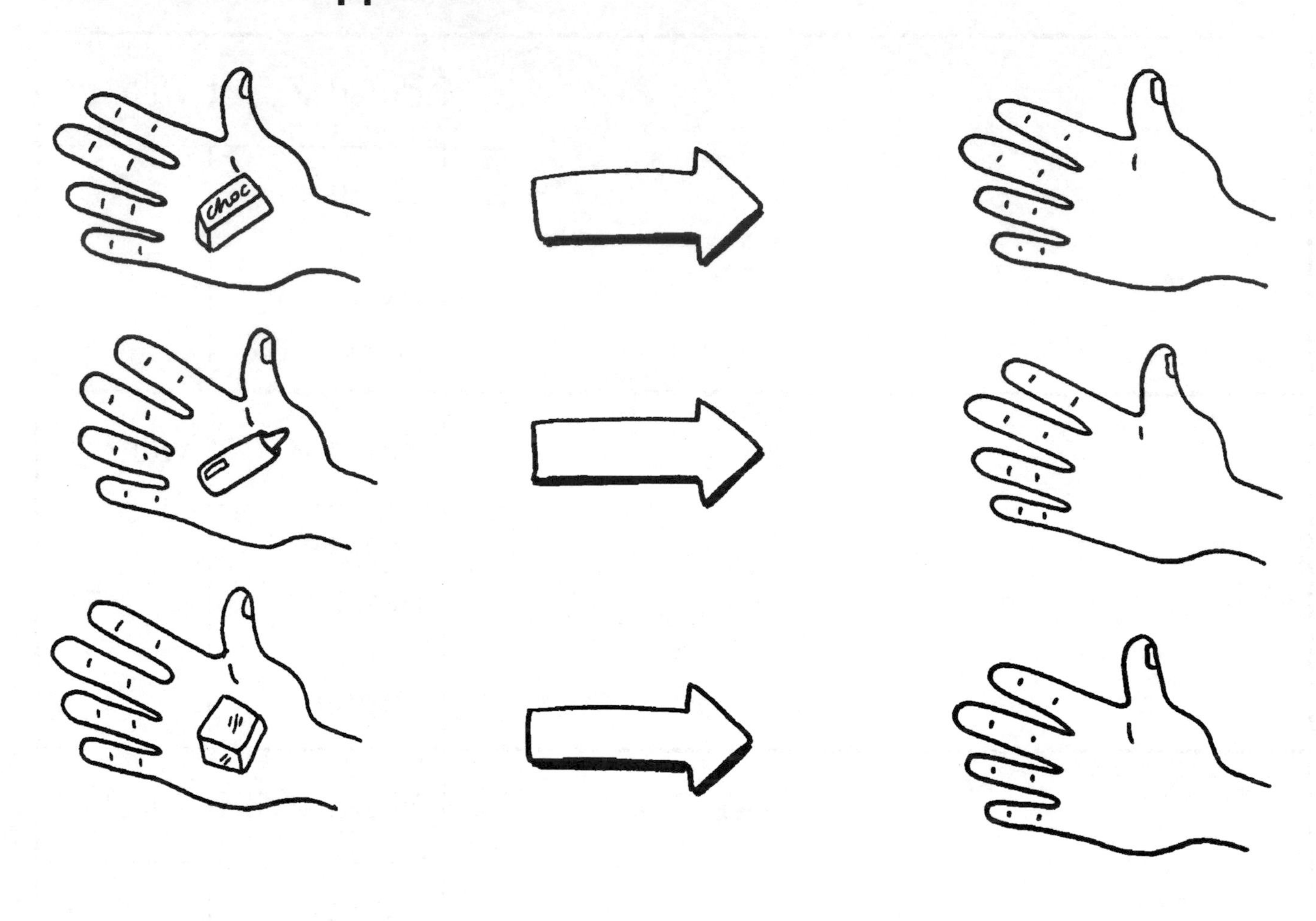


Which things changed?

Why do you think they changed?

# Changing for good?

Name ______________

Date ______________

| My clay model before firing | My clay model after firing |
| --- | --- |
| How it feels<br>1 ________ 2 ________ 3 ________ | How it feels<br>1 ________ 2 ________ 3 ________ |

**Word box**

soft rough cold hard squashy
wet warm smooth dry

**Colour the star red if you think the model has changed for good. Colour the star blue if you think it hasn't.**

# The wise man and the foolish man

Name ______________________

Date ______________________

Build a house with 10 bricks on stone.

Build a house with 10 bricks on sand.

Use a watering can to make it rain.

What happened?

What happened?

How did you make this test fair?

# Ice balloons

Name ______________________

Date ______________________

**Write or draw what you have found out about ice balloons.**

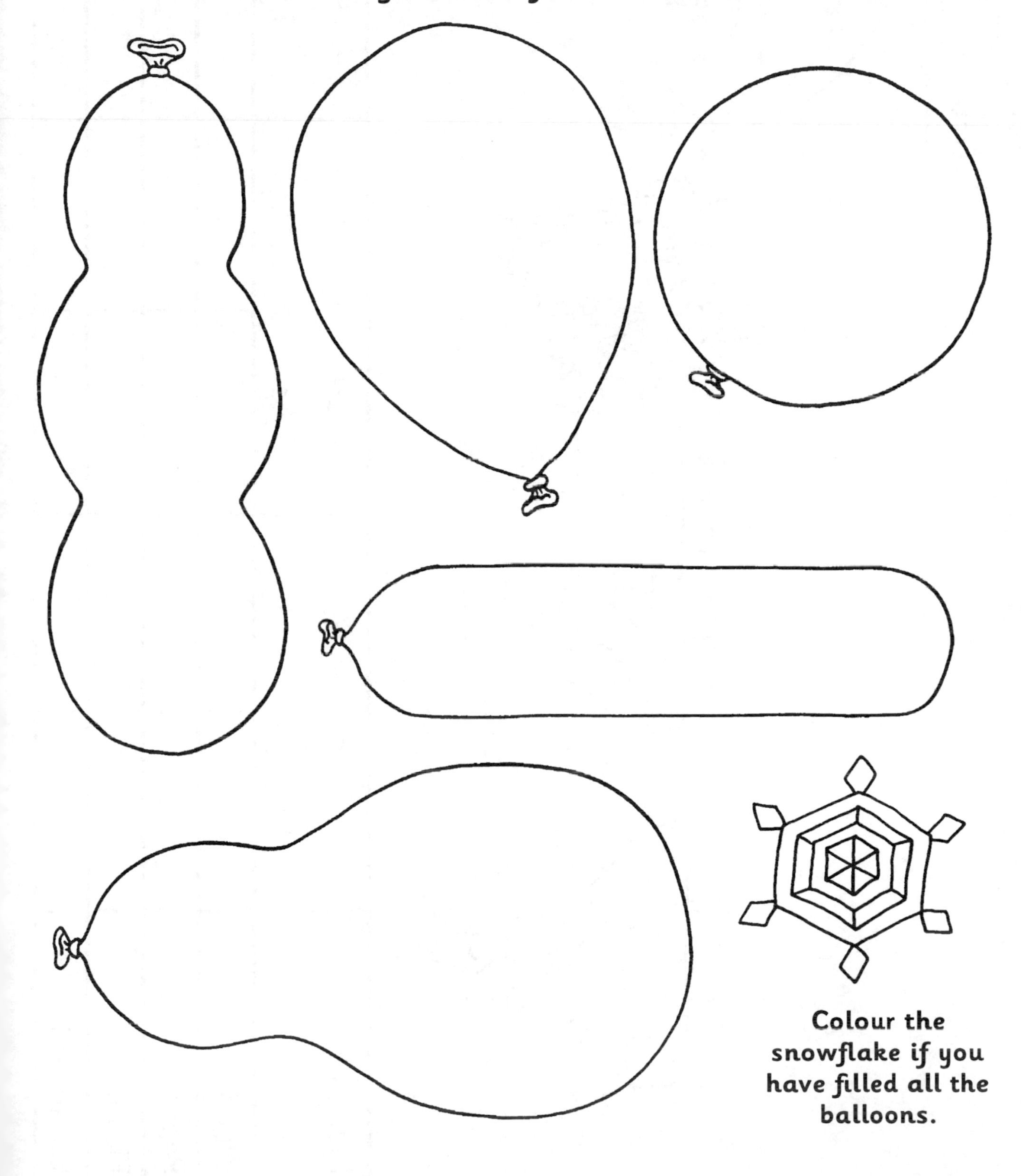

**Colour the snowflake if you have filled all the balloons.**

# Testing the bridge

Name ______________

Date ______________

| Materials | Bends ✓ ✗ | | Breaks ✓ ✗ | | Rigid ✓ ✗ | |
|---|---|---|---|---|---|---|
| | 1 kg | 5 kg | 1 kg | 5 kg | 1 kg | 5 kg |
| Wood | | | | | | |
| Paper | | | | | | |
| Plastic | | | | | | |
| Cardboard | | | | | | |

# Electricity in the home

Name ______________

Date ______________

**Colour red all the electrical dangers in this kitchen.**

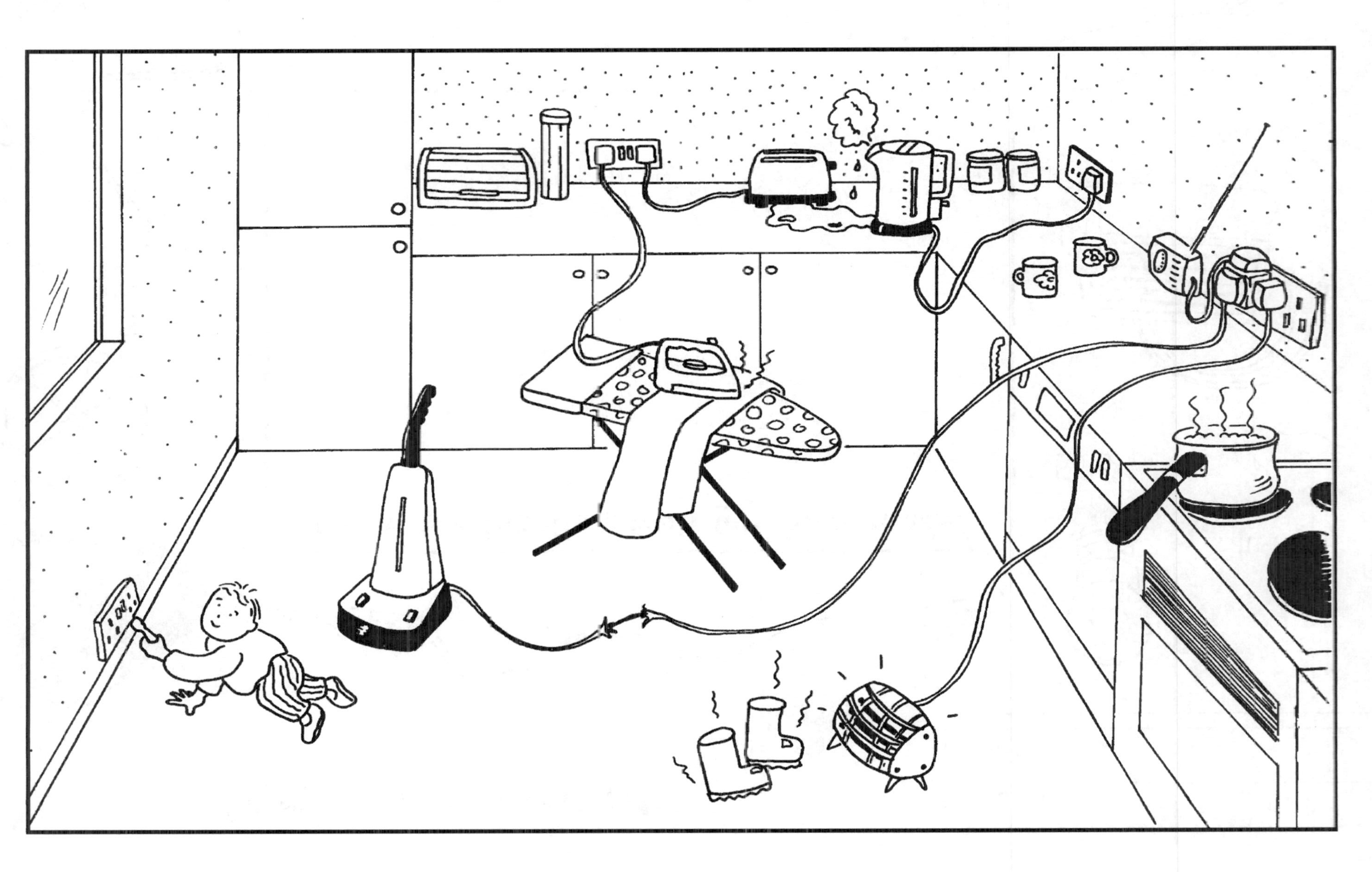

# Making a circuit

Name ______________

Date ______________

Draw and label a circuit you have made.

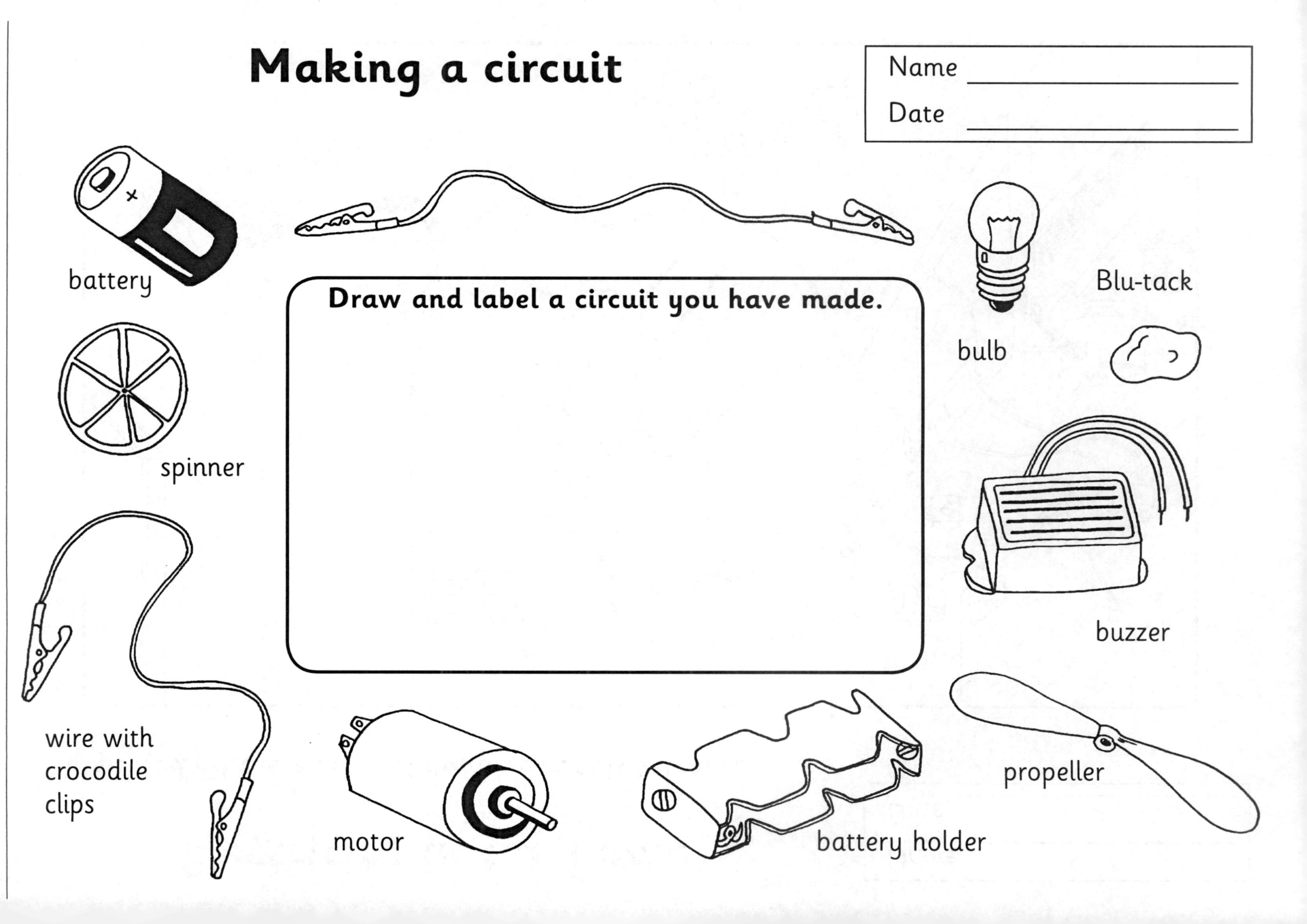

# Which vehicle goes furthest?

Name ______________________

Date ______________________

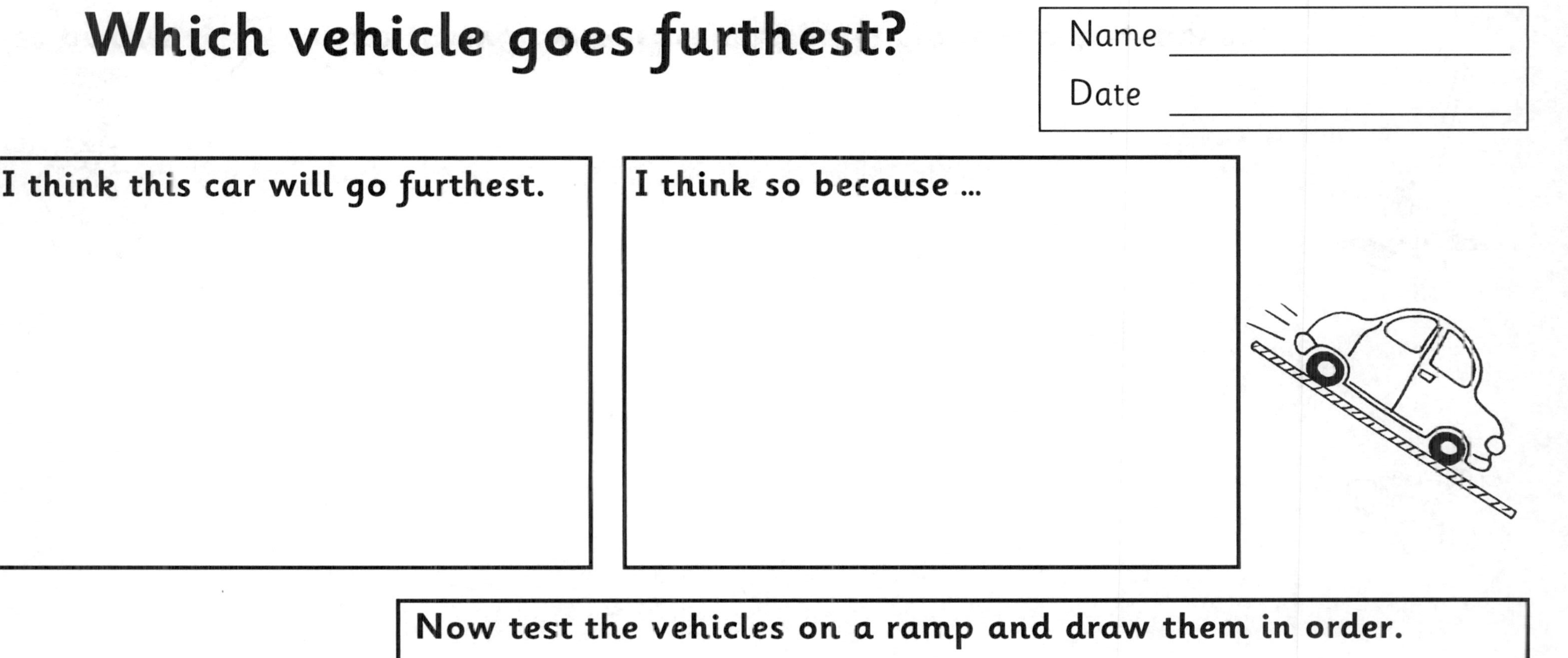

I think this car will go furthest.

I think so because ...

Now test the vehicles on a ramp and draw them in order.

Was your guess right? Yes ☐ No ☐

# Moving toys

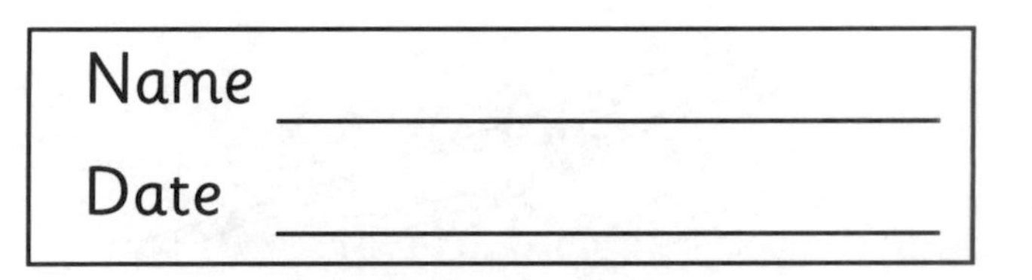

Choose a moving toy to sketch.

Use arrows to show how pushes and pulls make this toy move.

# Investigating sails

Name ______________________

Date ______________________

**Make three different shaped sails. Draw them on the boats.**

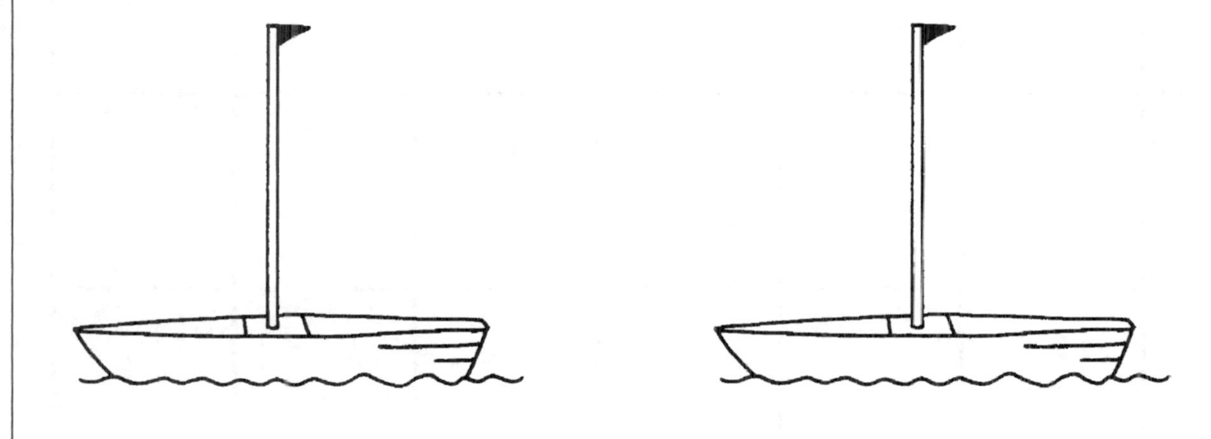

I think the ______________ sail will go furthest because ______________

______________________________________________

What happened? ______________________________

Was your test fair? Yes ☐ No ☐

Why? ________________________________________

# Investigating with forces

Name ____________________

Date ____________________

| Test these things. Put a ✓ or a ✗. | (hair dryer) | | (air pump) | |
|---|---|---|---|---|
| | slowed down | changed direction | slowed down | changed direction |
| beach ball | | | | |
| toy car | | | | |
| golf ball | | | | |
| cardboard tube | | | | |
| marble | | | | |
| tin can | | | | |
| | | | | |

# Floating and sinking

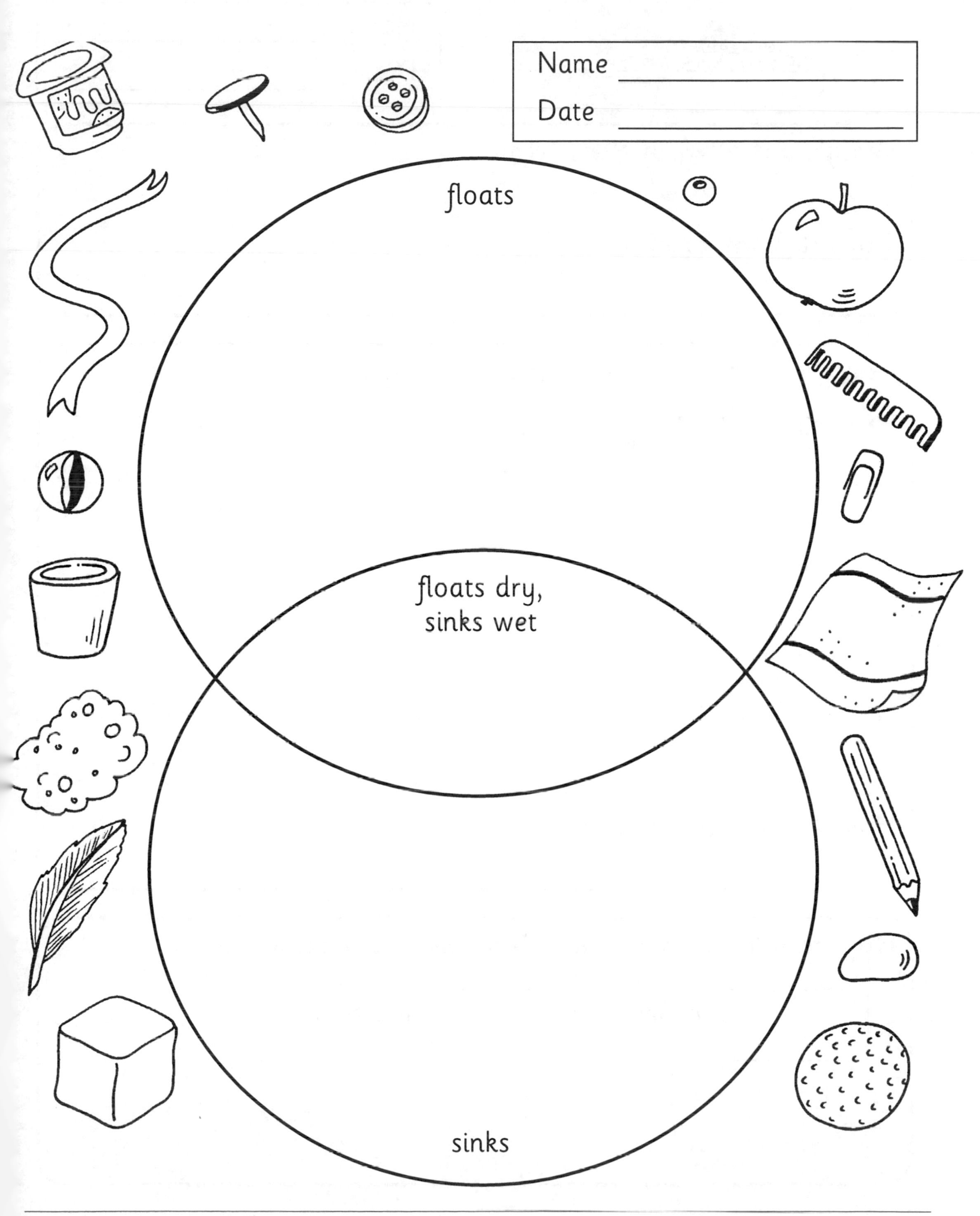

# Boats and cargo

Name ______________________

Date ______________________

**Weigh out 1 kg of Plasticine.**

**Does it float or sink?**

**Can you make it into a floating shape?**

**Draw your boat.**

**How long is your boat?** ______________________

**How much cargo can you load before it sinks?**

[ ] cubes [ ] marbles

# Light sources

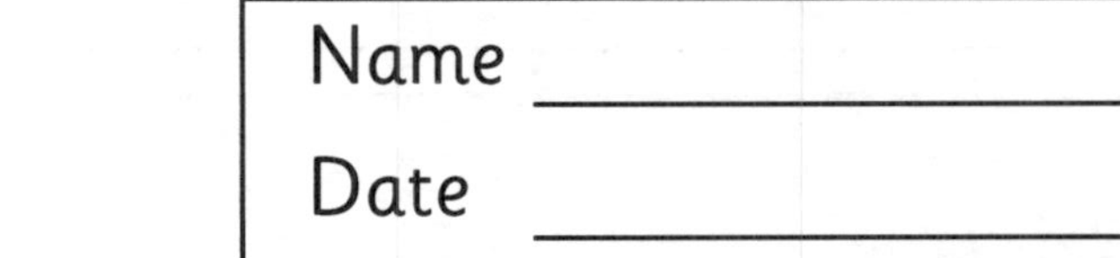

Name ____________

Date ____________

**Finish the picture by drawing in as many light sources as you can.**

# Shadows

Name ______________________

Date ______________________

Draw the shadows.

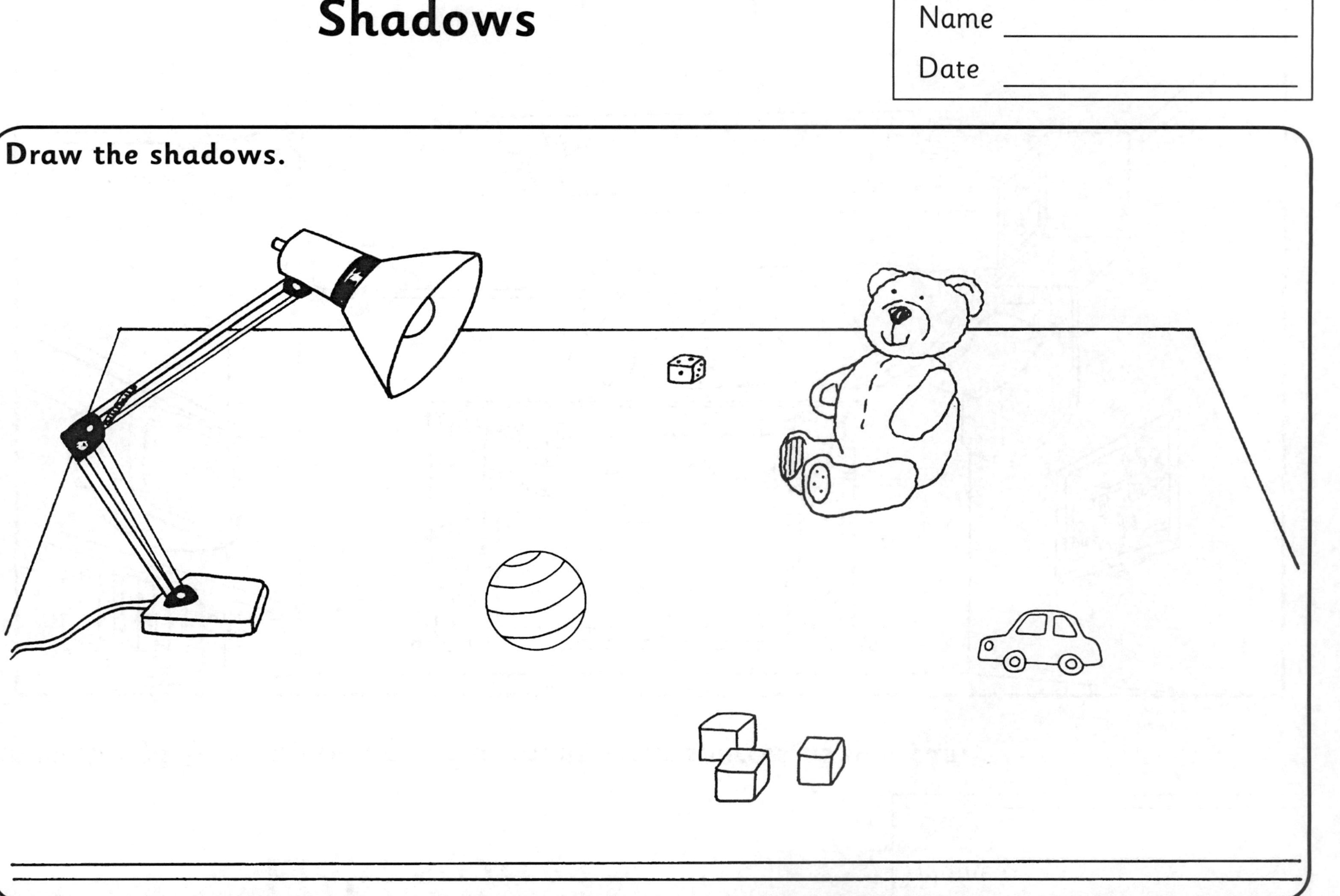

# Colours in the dark

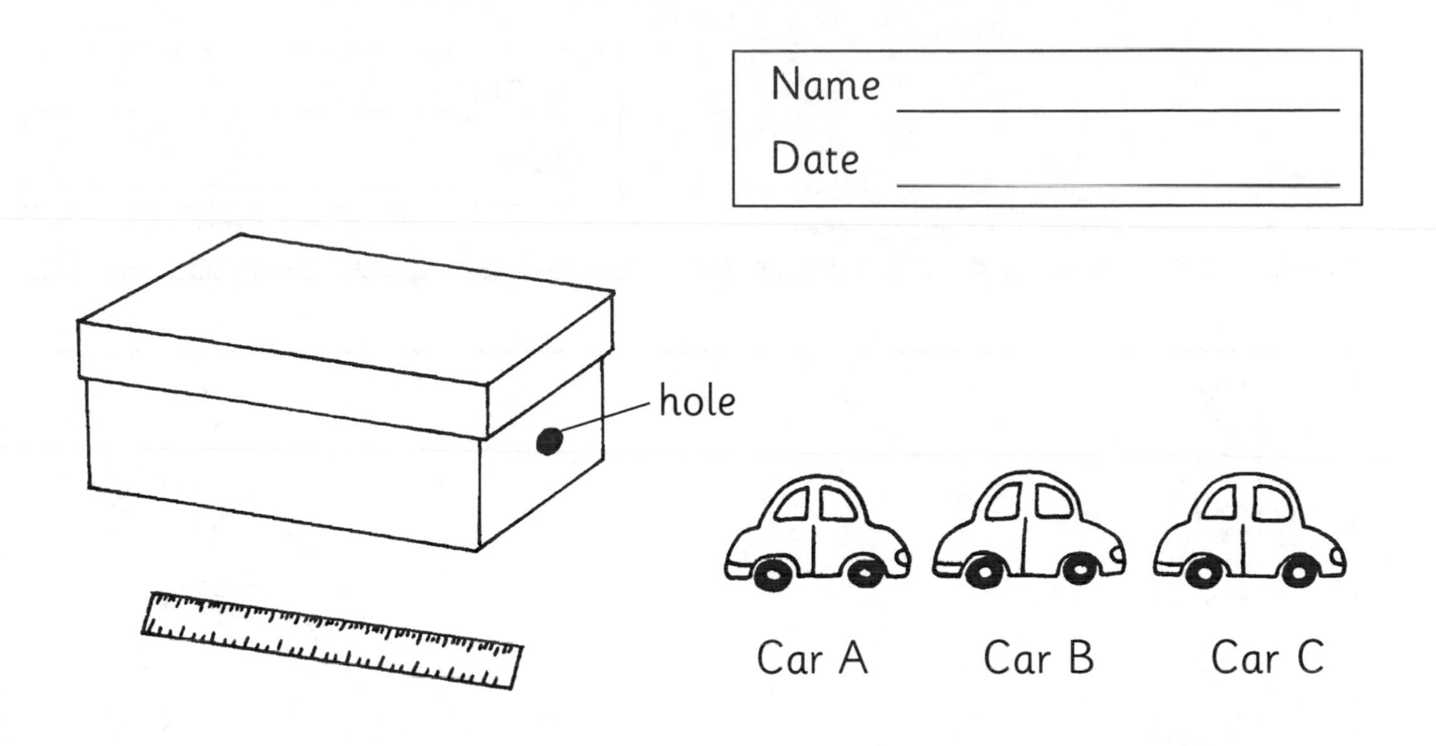


**Work with a friend. Put Car A in the shoe box. Look through the hole. Put a tick 3 in the box if you can tell what colour it is. Do the same with the other cars.**

A ☐  B ☐  C ☐

**Now try again. This time slowly raise the lid until your friend can tell the colour. Measure how far open the lid is for each car.**

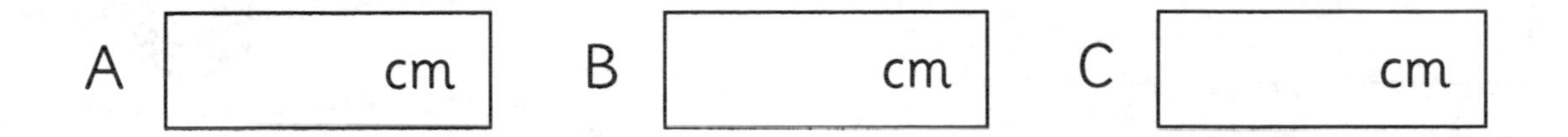


**Which colour shows up best in the dark?**

# Sounds walk

| Name | ______________ |
|---|---|
| Date | ______________ |

**Draw or write what sounds you heard on your sounds walk.**

# Making sounds

Name ____________

Date ____________

## How is the sound made?

# Investigating sound

Name ____________________

Date ____________________

**How far away can you hear these sounds?**

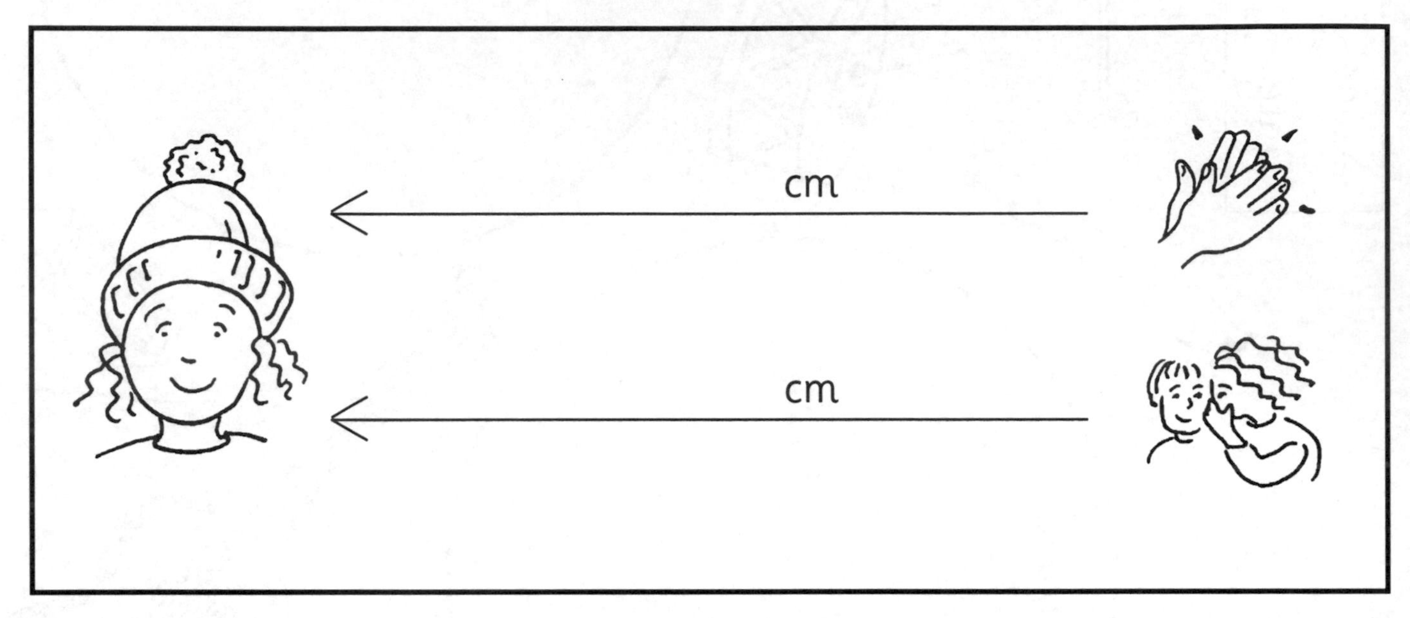


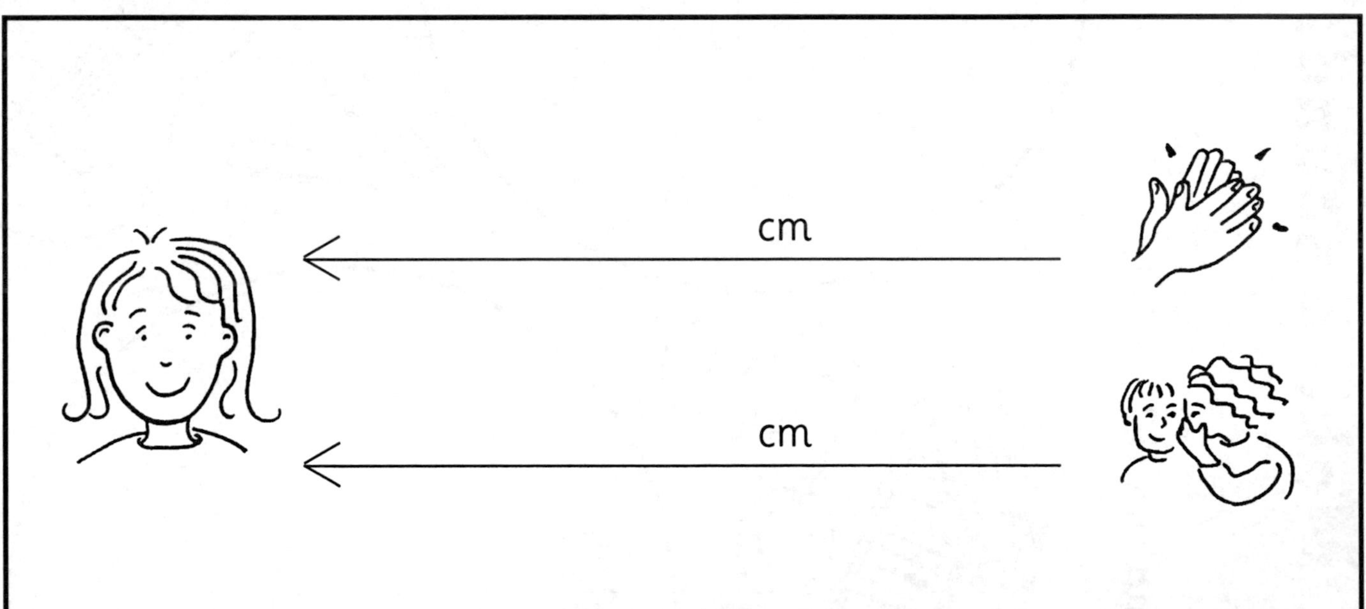


**How could you hear the sounds better?**

Lightning Source UK Ltd.
Milton Keynes UK
UKOW03f0007240915

259118UK00003B/11/P